This Notebook Belongs To

SCAN ME

Made in the USA
Monee, IL
13 August 2022

11532130R00065